P9-DND-072

The Edge *of the* Sky

The Edge *of the* Sky

All You Need to Know About *the* All-There-Is

Roberto Trotta

BASIC BOOKS
A Member of the Perseus Books Group
New York

Copyright © 2014 by Roberto Trotta
Published by Basic Books,
A Member of the Perseus Books Group

All rights reserved. Printed in the United States of America. No part of this book may be reproduced in any manner whatsoever without written permission except in the case of brief quotations embodied in critical articles and reviews. For information, address Basic Books, 250 West 57th Street, 15th Floor, New York, NY 10107.

Books published by Basic Books are available at special discounts for bulk purchases in the United States by corporations, institutions, and other organizations. For more information, please contact the Special Markets Department at the Perseus Books Group, 2300 Chestnut Street, Suite 200, Philadelphia, PA 19103, or call (800) 810-4145, ext. 5000, or e-mail special.markets@perseusbooks.com.

Illustrations by Antoine Déprez
Designed by Jeff Williams

Library of Congress Cataloging-in-Publication Data

Trotta, Roberto.
The edge of the sky : all you need to know about the all-there-is : (using only the ten hundred most-used words in our tongue) / Roberto Trotta.
pages cm
ISBN 978-0-465-04471-9 (hardcover)
ISBN 978-0-465-04490-0 (e-book)
1. Cosmology—Popular works. 2. Astronomy—Popular works. I. Title.

QB982.T76 2014
520—dc23
2014020067

10 9 8 7 6 5 4 3 2 1

To my two E's

What's Inside

Before You Read This Book

This book came out of a little idea: that it should be possible to talk about very hard things in a straight-forward way that all people can understand.

A way to do that is by talking with only the most-used ten hundred words in our tongue. When I heard about this, I thought that it could be fun to use it to explain the entire All-There-Is.

You will find in this book a short story about what we think the All-There-Is is made of, and how it got to be the way it is. You will also see that there is so much more that we don't yet understand.

And we'll even try to glance past the edge of the sky and imagine what might lie there.

All of this using only the most-used ten hundred words. If you need a little help with some of the new expressions I've used, you'll find them explained with normal words at the end of the book.

I hope you enjoy it.

About the Guy Who Wrote This Book

I am a Doctor at a college where we help students to understand how the world works. People call me a Doctor, but please don't come to me with your broken arm or a bad cold. I'm not that kind of doctor!

My college is in one of the most busy and beautiful cities in the world, which has a water road running through it and guards in red jackets and black head-covers. The head of state is a dear old lady who comes from a line that once had great power, and people in the street wave when they see her at her window.

I ask questions about how everything around us works, and why it is the way it appears to be. I also love to talk to people who have a different job from mine, and they often have great questions to which I don't have any answers.

I have written this book to tell you some of the things we have learned, and the many things we still don't know.

If you want to talk to me, please look for me on the world-wide computer or send me a note with the little blue flying animal. I would love to hear from you.

The Ten Hundred Most-Used Words in Our Tongue

a able about above accept across act actually add admit afraid after afternoon again against age ago agree ah ahead air all al-low almost alone along already alright also although always am amaze an and anger angry animal annoy another answer any anymore anyone anything anyway apartment apparently appear approach are area aren't arm around arrive as ask asleep ass at attack attempt attention aunt avoid away baby back bad bag ball band bar barely bathroom be beat beautiful became because become bed bedroom been before began begin behind believe bell beside besides best better between big bit bite black blink block blonde blood blue blush body book bore both bother bottle bottom box boy boyfriend brain break breakfast breath breathe bright bring broke broken brother brought brown brush build burn burst bus business busy but buy by call calm came can can't car card care carefully carry case cat catch caught cause cell chair chance change chase check cheek chest child children chuckle city class clean clear climb close clothes coffee cold college color come comment complete completely computer concern confuse consider continue control conversation cool corner couch could couldn't counter couple course cover crack crazy cross crowd cry cup cut cute dad damn dance dark date daughter day dead deal dear death decide deep definitely desk did didn't die different dinner direction disappear do doctor does doesn't dog don't done door doubt down drag draw dream dress drink drive drop drove dry during each ear early easily easy eat edge either else empty end enjoy enough enter entire escape especially even evening eventually ever every everyone everything exactly except excite exclaim excuse expect explain expression eye eyebrow face fact fall family far fast father fault favorite fear feel feet fell felt few field fight figure fill finally find fine finger finish fire first fit five fix flash flip floor fly focus follow food foot for force forget form forward found four free friend from front frown fuck full fun funny further game gasp gave gaze gently get giggle girl girlfriend give given glad glance glare glass go God gone gonna good got gotten grab gray great green greet grin grip groan ground group grow guard guess gun guy had hadn't hair half hall hallway hand handle hang happen happy hard has hate have haven't he he'd he's head hear heard heart heavy held hell hello help her here herself hey hi hide high him himself his hit hold home hope horse hospital hot hour house how however hug huge huh human hundred hung hurry hurt I I'd I'll I'm I've ice idea if ignore imagine immediately important in inside instead interest interrupt into is isn't it it's its jacket jeans jerk job join joke jump just keep kept key kick kid kill kind kiss kitchen knee knew knock know known lady land large last late laugh lay lead lean learn least leave led left leg less let letter lie life lift light like line lip listen little live lock locker long look lose lost lot loud love low lunch mad made make man manage many mark marry match matter may maybe me mean meant meet memory men mention met middle might mind mine minute mirror miss mom moment money month mood more morning most mother mouth move movie Mr. Mrs. much mum mumble music must mutter my myself name near nearly neck need nervous never new next nice night no nod noise none normal nose not note nothing notice now number obviously of off offer office often oh okay old on once one only onto open or order other our out outside over own pack pain paint pair pants paper parents park part party pass past pause pay people perfect perhaps person phone pick picture piece pink piss place plan play please pocket point police pop position possible power practically present press pretend pretty probably problem promise pull punch push put question quick quickly quiet quietly quite race rain raise ran rang rather reach read ready real realize really reason recognize red relationship relax remain remember remind repeat reply respond rest return ride right ring road rock roll room rose round rub run rush sad safe said same sat save saw say scare school scream search seat second see seem seen self send sense sent serious seriously set settle seven several shadow shake share she she'd she's shift shirt shit shock shoe shook shop short shot should shoulder shouldn't shout shove show shower shrug shut sick side sigh sight sign silence silent simply since single sir sister sit situation six skin sky slam sleep slightly slip slow slowly small smell smile smirk smoke snap so soft softly some somehow someone something sometimes somewhere son song soon sorry sort sound space speak spend spent spoke spot stair stand star stare start state stay step stick still stomach stood stop store story straight strange street strong struggle stuck student study stuff stupid such suck sudden suddenly suggest summer sun suppose sure surprise surround sweet table take taken talk tall teacher team tear teeth television tell ten than thank that that's the their them themselves then there there's these they they'd they're thick thing think third this those though thought three threw throat through throw tie tight time tiny tire to today together told tomorrow tone tongue tonight too took top totally touch toward town track trail train tree trip trouble true trust truth try turn twenty two type uncle under understand until up upon us use usual usually very visit voice wait wake walk wall want warm warn was wasn't watch water wave way we we'll we're we've wear week weird well went were weren't wet what what's whatever when where whether which while whisper white who whole why wide wife will wind window wipe wish with within without woke woman women won't wonder wood word wore work world worry worse would wouldn't wow wrap write wrong yeah year yell yes yet you you'd you'll you're you've young your yourself

The Edge *of the* Sky

Part 1

The First Night

In the clear night
Her dark hair
Mirrors the stars

She steps out and takes a deep breath. The air is cold, and a little white ice shadow comes out of her mouth.

"This is harder than I thought," she whispers to herself.

The guy back home, the one who sent her there, did warn her. "The first time is the hardest. Every person says so," he said.

It's true. Just getting here has been hell. First, it was two long trips in those big flying cars. Then another jump, this time in a much smaller flying car. For the last leg, it was only her and the guy driving the tiny flying car, in fact.

A guy from the office picked her up with his big car, one that you can drive to almost any place you like without needing a road. She immediately knew he was from the office because he wore on his jacket the sign that all people working there have. Even the car had it on its doors.

It took them many hours on a road that became smaller and smaller the further up they went. The guy who was driving didn't like to talk much, so she looked out of the window in silence. The trees quickly gave way to nothing. After a while, nothing man-made was in sight. And the road went up, and up, and up.

When they finally arrived, she took five steps and almost fell. Her head felt light, her stomach was not right, and it seemed to her that she had just drunk ten bottles of red. She was expecting it, of course, but was surprised all the same.

That was a few days ago. But tonight her time has come. She has spent so many years getting ready for this that she almost can't believe it's happening. She takes a last look around her: the night is beautiful, and this place sits so high up that there are no lights at all in sight.

Except for those above her.

Those lights are why she is here. There are so many of them in this dark night, in this place so far away from everything else, that she feels like falling toward them every time she looks at the sky.

"We have forgotten what the night sky really looks like," she thinks as she goes back inside.

"We have forgotten what the night sky really looks like," she thinks as she goes back inside.

The room in which Big-Seer sits is as cold as the outside: warming it up wouldn't make things easier for her, since the eye of Big-Seer would not see right.

From her desk she can see Big-Seer sitting quietly in his room on the other side of a large window, his big eye closed. He is waiting. A hundred hundred colored lights blink in front of her. She closes her eyes for a second and relaxes, and then with fast touches of her hands she sets off to work.

She knows that Big-Seer is ready. She has spent the last few nights making sure that he was: she carefully checked his position and, most of all, the sight in his big eye. She used the blinking lights to fix it in just the right way. It is long, hard work: his glass eye is so big that it would take ten people and another six holding hands to go around it.

Any fault, however tiny, could stop Big-Seer from doing his job. She knows this, and that's why she was nervous. But now everything is ready, and she is calm and focused.

"Now is the time," she says to herself.

No one can hear her, for she is alone in the room. With the touch of a finger, she pushes the last red light and with barely a noise Big-Seer opens his big eye. The light from the stars falls on her arm.

All of a sudden, Big-Seer jumps to life. He moves his glass gaze slowly but without pausing. He knows what he is looking for.

"Dark matter, here we come!" she exclaims.

Part 2

The Crazy Stars

Blue worlds
Going around
Points of light

As Big-Seer turns his huge eye to the sky, she sits back and thinks of how much our ideas about all that is out there have changed over the years.

The old people thought that we were right in the middle of all there is.

Everything else in the All-There-Is went around our Home-World: every evening, the stars would come up in the sky, move across it, and disappear in the morning. The Sun would chase them away. At night, the Sun's Sister would appear: at times, she would be full and round. Night after night, she would gently disappear, until, after two weeks, she would become dark. Then she would grow to become full and bright again.

The stars formed animals and gods in the sky, as well as huge people.

The people, animals, and gods were different in the warm and cold months of the year. They brought rain and warm times, ice and strong waters. They made your food grow and killed your animals. They decided whether your son or daughter would be sick. They told you whether it was the right time to build a new house or to put a food-tree in the ground.

They decided your life.

But the gods with the most power, besides the Sun and his Sister, were the stars that did not follow the others.

These Crazy Stars did not need to go around in the same way as all the other stars. They were free to go in the same direction for a while, then stop, turn around, and go back. And they were brighter than most of the other stars.

People feared them, and watched them carefully. The Crazy Stars had big names and great power.

He-Who-Talks-for-the-Gods could run faster than every other Crazy Star. He was always moving here and there. When you died, it was his job to show you around the world of the dead.

The stars formed animals and gods in the sky, as well as huge people.

The She-God of Love could make you fall in a second for a man or woman you were passing in the street. She was beautiful, and she rose from the waters perfect and soft-skinned.

The Fight-God was young, strong, and crazy. He could attack any person he wanted, and did not need a reason to do so. The third month of the year is still named after him.

The Head-God was the strongest of them all. He was the father of all the Crazy Stars, and they had to do as he wanted. He could send fire to hit you and always had a great flying animal with him.

The Great-Father-God was the father of the Head-God. He could control time, and when the new year arrived, a big party was given for him.

That was all there was then.

The All-There-Is was all built around us. Easy. Ordered. And wrong.

"If the old people were around today, they would have a really hard time naming all the new Crazy Stars we have found!" she thinks, as her hands fly from one red light to the other.

On top of the ones the old people knew and feared, people who lived not so long ago found three more Crazy Stars.

By then, it was realized that the Crazy Stars do not go around our Home-World, as the old people thought. The

Crazy Stars actually go around the Sun, and so does our Home-World. They do so in a straight-forward way, which is actually not as crazy as it seemed. It is the marrying of the dances of the Crazy Stars and of our own Home-World around the Sun that makes the Crazy Stars appear to go around in such a strange way.

The three new Crazy Stars are much harder to see because they are not as bright as the ones the old people knew.

To spot them in the sky, you need a Far-Seer, a little brother of Big-Seer. A Far-Seer works in the same way as Big-Seer does, but can only see so far.

These new Crazy Stars were named after some other old gods: the Water-God, the Sky-God, and the Hell-God. If we add our Home-World, there are three times three Crazy Stars going around the Sun. A few years ago, it was decided that the Hell-God is actually too small to be a Crazy Star like all the others. Now it is called a minute Crazy Star.

We have looked at all those Crazy Stars using Far-Seers and Big-Seers.

We have sent flying Far-Seers to visit all of them and to take pictures from close by. We have landed space-cars on some of them, and put people on the Sun's sister, and they have driven around looking for small life. We haven't found any.

After having looked at all the Crazy Stars going around the Sun, we have found that we are the only form of life that knows what it means to live.

But a few years ago, student-people found that there might be many, many more places where people could live out there.

Student-people are different from other people. They spend their entire life asking questions, and as soon as they have found out the answers, they start all over again with new, harder questions.

They love questions. And they never run out of them. The best questions for student-people are the ones that begin with Why? These questions are also the hardest.

When a student-person finds a good answer to a hard question, the other student-people will gasp, hug each other, and then throw a party. Those parties never last long, for student-people are in a hurry to go back to work and find new answers. The best party for student-people is one that takes place every year in a cold place with lots of ice-water, close to the top end of our Home-World. It is named after the person who set it up over a hundred years ago, Mr. Nobel. All student-people would like to be asked to go to this party, but only up to three can go together each year.

A few years ago, two student-people found the first Crazy Star that was not going around our Sun.

We now know that there are Crazy Stars going around most of the stars we see in the sky, yet no one before had figured out how to see them.

This is because far-away Crazy Stars are not bright enough to be seen, even with the biggest Big-Seer. They are little points of light going around a much brighter star. The light from the star is so much brighter that it makes them disappear.

But the two student-people found that they could tell that a Crazy Star was going around a far-away star, even if they could not see the Crazy Star.

They realized that the star moves a little when the tiny Crazy Star passes in front of it, and pulls the star this way or that way. So by looking at the small changes in where the star was in the sky, they figured out that a Crazy Star was going around it.

Student-people have now found hundreds and hundreds of Crazy Stars going around far-away stars. Most of these Crazy Stars are big, like the Head-God, and very different from our Home-World.

For life as we know it to live on far-away Crazy Stars, they need to be close enough to their star (or it would be very cold), but not too much (or it would be too hot).

We haven't found Crazy Stars like our Home-World yet, but it's only a question of time. Every day, student-people come up with new, better ways of looking for even smaller Crazy Stars, and soon they will find one that could have water and trees and animals and perhaps even people living on it.

Even if that happens, those new Home-Worlds will be too far away for us to ever visit.

Just sending a card there to say hello would take years, even if the card was flying as fast as light. That's the fastest letter there could be! Sending people there would be even harder, and we do not know how to do it yet.

But to know that there is life somewhere else out there would be shocking news. It would change the way we think about us and our place in the All-There-Is.

Part 3

A Bigger Place

Behind the stars
Space-time grows
Silent

The colored lights in front of the student-woman flash quietly on and off. Big-Seer goes about his job in silence. Nothing breaks the perfect calm of the dark night.

It will take at least another two hours before Big-Seer has finished. This is because Big-Seer is looking at a far-away Star-Crowd, from which very little light reaches us.

When the light set off on its long trip to us, our Home-World had not been put together yet. That's how long it took for the light to get here—even though light can go around our Home-World seven times in one second!

It is hard to picture how big the All-There-Is actually is. A student-man called Doctor Hubble was the first to understand it.

About a hundred years ago, Mr. Hubble was using one of the biggest Far-Seers at the time. He was working at a student-people place near a city that, a few years later, would become the most important place for people making movies.

He was using the Far-Seer to look at some White Shadows in the sky that appeared to be made of some sort of smoke. They were very hard to make out, and no one knew what they were.

Part of the problem was that no one knew how far away they were. If they were quite near to our Home-World, they could not be too large. But if instead they were far away, they would have to be huge to appear that small even with the finest Far-Seer people could build.

Mr. Hubble set his mind to sort this out.

If he could answer the question of how far away the White Shadows were, he reasoned, he would find out whether they were part of our own Star-Crowd. And if they weren't, this would show that the All-There-Is was much bigger than anyone thought.

Most student-people at the time believed that all there was in the All-There-Is were the stars we could see in our Star-Crowd, of which the Sun is part.

When you look at the sky on a dark night, the group of stars near us looks like a white road running right across the sky. So we call it the White Road.

The White Road is made of many, many stars, most of which are too small to be seen without a Far-Seer. It's hard to imagine how many stars are in the White Road.

If you tried to number them, and you could get a hundred done every second, it would still take you a hundred years to go through them all!

So the question that many student-people were asking was whether the White Road was all there is in the All-There-Is. And if it wasn't, what was further out, where the White Road stopped? Was it black, empty space going on without end, or was there something else?

Doctor Einstein was to become one of the most important student-people ever. He had a quick brain and he had been thinking carefully about the building blocks of the All-There-Is. To his surprise, he found that light was the key to understanding how far-away things in the sky—Crazy Stars, our Star-Crowd, and perhaps even the White Shadows—appear to us.

He began by asking what he would see if he could fly as fast as light and someone else was coming toward him. He knew that the answer was strange: it didn't matter whether the other person was still or moving as fast as light. Mr. Einstein would always see the other person approach as fast as light. This is not what you would see if you drove a car toward another car: how fast the other car comes at you changes if the other car is moving or not.

Mr. Einstein knew that things were different for light. Two student-people called Doctor Michelson and Doctor Morley had tried a few years before to time light as it flew toward two mirrors, one of which was moving while the other was still. They found that it didn't matter which mirror the light was flying to; the time it took to get there and back was always the same.

You could not explain this using the normal idea of space and time. Mr. Einstein then said that space and time had to be married and form a new thing that he called space-time. Thanks to space-time, he found that time slows down if you fly almost as fast as light and that your arm appears shorter in the direction you are going.

He then asked himself what would happen if you put some heavy stuff, as heavy as a star, in the middle of space-time. He was the first to understand that matter pulls in space-time and changes the way it looks. In turn, the form of space-time is what moves matter one way or another.

It followed that light from stars and the White Shadows in the sky would also be dragged around by the form of space-time. Understanding space-time meant understanding where exactly and how far away from us things are in the sky.

As an idea, this was very different from what Mr. Newton had said a long time ago. For Mr. Newton, space and time did not talk to each other, never married, and lived different lives.

Mr. Einstein's idea seemed weird at first, and it was hard to understand, even for student-people. What was needed was a way to tell who was right, Mr. Einstein or Mr. Newton.

It was no easy matter, and something really strange was needed: night during day time. This happens when the Sun's Sister passes right in front of the Sun so that the two are exactly lined up. For a few minutes, and only in a given place, the day becomes night and all goes dark. Animals go quiet and people gasp.

At this moment, the sky becomes dark and you can see the stars, even if it's during the day. Thanks to his new idea, Mr. Einstein had said that the position of the stars around the Sun would not be the same as usual.

This is because the Sun is very heavy and so would drag space-time around. Space-time then would pull the light from the stars away from a straight line. To us, it would look as if the stars were somewhere else.

To tell whether Mr. Einstein's idea was right, student-people had to look at those stars around the Sun with a Far-Seer to see if they were in the place Mr. Einstein said they would appear to be.

And they were!

Mr. Einstein's idea about space-time (weird as it sounded) was actually right.

Mr. Einstein then began to wonder what would happen if he used his space-time idea for the entire All-There-Is.

He quickly understood that the All-There-Is could not stay still, because space-time would be forced to change by the matter we see all around us, like other stars, the White Road, and perhaps the White Shadows (if they were big enough, which no one knew).

Other student-people had imagined, starting from Mr. Einstein's space-time idea, that the All-There-Is might be growing with time. But Mr. Einstein did not want to believe this. He thought that the All-There-Is had been the same for all time. It had no start and would have no end.

What he needed to do, Mr. Einstein decided, was to add something else to his space-time idea so that the All-There-Is would not change over time.

At this moment, the sky becomes dark and you can see the stars,
even if it's during the day.

So he imagined a new force that was all around us, in space. Where this force was coming from no one could see, not even with the biggest Far-Seers: this was the idea of the Dark Push.

By pushing away space-time, the Dark Push would act against the pull of matter.

This way, the pull and the push would stop each other just so, and the All-There-Is would never change. It would be like a person standing on her head, always just a hair away from falling toward one side or another.

That was the situation when Mr. Hubble started looking at the White Shadows in the sky. Inside each White Shadow, he looked for a type of star that changes over time. The light from those stars gets brighter for a while, then goes back to normal, then grows brighter again, and so on. The time it takes for the light to go back to normal tells you how far away they are. In this way, he found that the White Shadows are very far away, and definitely outside the White Road.

We now know that each of those White Shadows is actually as big as our own White Road, and each is made of as many stars, or even more. We call them Star-Crowds.

But there was more: Mr. Hubble found that the color of the light we get from the White Shadows was redder the further away the White Shadows were. We say that light gets tired during the trip—the more tired, the further away it comes from.

That light gets tired on its way to us was something that student-people could explain using Mr. Einstein's space-time idea.

Imagine that space-time between us and the far-away White Shadows is changing with time, and becoming bigger. Light is made of waves, and as it rides through a growing space-time, those waves are drawn away from each other and get longer and longer. The longer the waves, the redder the light looks—more so the further away the White Shadows are, exactly as Mr. Hubble found. Student-people call this "red-shift."

Mr. Hubble also found that it didn't matter where on the sky the White Shadows were—light was getting tired in the same way from all directions. The All-There-Is is getting bigger all around us, and this makes it look like all White Shadows are moving away from us.

But this does not mean that we are in the middle of the All-There-Is, for that would be really strange!

Seen from other White Shadows, our own White Road is also moving away from everything else.

Mr. Hubble's work showed that Mr. Einstein's first idea was right. The All-There-Is is changing with time, and Mr. Einstein had been wrong in adding in the Dark Push to stop this from happening.

Later, Mr. Einstein said that the Dark Push had been a stupid idea.

That the All-There-Is was growing meant another jump in our understanding of where it all comes from. Going back in time, the All-There-Is was smaller. At some point in time, it must have been smaller than the White Road now. Earlier still, it would have been so small as to be a tiny point. And that was when everything began.

We call this moment the Big Flash.

Part 4

The Big Flash

Hot kisses
Leave light
Very tired

She would never have thought to end up here.

She had not been one of those kids with a clear idea of what they are going to become. And to become a student-woman was not something she had dreamed of—even less to become a student-woman who studies the All-There-Is.

Her family wanted for her a real job: a job everyone knew about.

A doctor—that was a great job. She would have been good at that, they thought. Or one of those people who wear horse hair on their head and try to trip up other people for a living. They explain how things have really gone, say, if someone has killed another person, and have to make sure they are believed.

But in school she found that what she really wanted to know was how things work, deep down. She kept asking Why? questions, and for every answer, there came an even bigger, deeper Why? question.

So she understood that what she really wanted to know was why the All-There-Is is the way it is, and how it works. That was the job of the student-people. She had to become one.

It had been hard but fun work. She loved finding better and better answers to her Why? questions, until there came a point when no one had the answers—yet.

This is why she is here now, with Big-Seer, looking for answers to a very hard question. "The real work has just begun," she says to herself, smiling.

The starting point of her question was the Big Flash. This was the moment when the All-There-Is began. Before this time, there was nothing—or perhaps there was, only we can't tell.

We can't say much about the moment of the Big Flash, but we have a good idea about what happened right after it.

This is really great, if you think how long ago the Big Flash happened! It was over one hundred times one hundred times one hundred times one hundred times one hundred years ago.

We know the age of the All-There-Is so well that it would be like being able to tell the day of the year a stranger in the street came to life to the nearest day just by looking at him.

We do know some other things about the Big Flash.

Right after the Big Flash, everything was very hot. The All-There-Is started from a single point, but then grew very, very quickly to become very, very large.

It is almost not possible to picture how fast it grew. Imagine breathing into a colored party ball, so that with every breath the ball becomes ten times bigger than before. If every breath took you an hour, you would have to keep going for over three days to make the ball grow as much as the All-There-Is grew right after the Big Flash. By that time, your party ball would have become much bigger than the White Road, so that one hundred party balls would fill the entire part of the All-There-Is we can see!

We don't know what made it grow so much, so fast. Student-people think that space was filled with something like the Dark Push—something that was pushing space-time away, making it bigger and brushing away any spots. This Early Push left space-time shaking with lots of tiny waves, which student-people think they have now picked up with a Far-Seer at the bottom end of our Home-World.

Whatever it was, much less than a second after the Big Flash the Early Push had disappeared and all sorts of tiny drops of matter as well as light came into being.

At the beginning, all the matter drops were hot and moved around quickly. Nearly every matter drop had a Sister Drop flying around, and when they met, they hugged each other and disappeared in a flash of light. All the drops would have gone and only light would be left over, if it wasn't for a strange fact.

Imagine a number of matter drops as large as the number of people who live in the land of Mr. Mao today. Each one of them had a matching Sister Drop, and when they found it, both disappeared.

Except for one.

Everything we see around us today is made of the few matter drops that did not have a Sister Drop and that escaped their death hug.

As space continued to grow bigger and bigger, it cooled down. During the next three minutes, when the left-over matter drops met another drop they liked, they kissed each other and stuck together. Most matter drops did not find any other drop to kiss, so they stayed alone. We call them the Single Drops.

Almost all the matter drops that kissed each other ended up as Heavier Drops, made of two pairs of different drops. Very few

Nearly every matter drop had a Sister Drop flying around, and when they met, they hugged each other and disappeared in a flash of light.

matter drops stuck together to form even bigger drops than the Heavier Drops.

At the end, there were about ten times as many Single Drops as Heavier Drops. Single Drops and Heavier Drops are the same kind of drops that today make up most of the Sun.

Also, a whole lot of much lighter Very Small Drops were still flying around like crazy.

After three minutes, the All-There-Is had grown too much for matter drops to kiss: they simply could not find each other any more in all that big, empty space!

Once matter drops stopped kissing one another, nothing much happened for a long time. The All-There-Is was still young, and stars hadn't yet come into being.

If you had been around at that time, you would not have been able to see anything much at all.

This is because light could not fly for long before hitting the Very Small Drops and getting kicked about. Light could not go in a straight line from one point to another.

It was like when you look out your window on a cold morning: there is ice on the road, and everything looks gray, and you can't see past your drive-way. This is because light gets kicked about by tiny water drops in the air before reaching your eyes.

Something like this was happening at that time, only everything was a hundred times hotter than on a hot summer's day.

After some more time, the All-There-Is had become cold enough for something interesting to happen.

Very Small Drops had been flying around all the time, hot enough to escape the hug of the Single Drops. But when the Very Small Drops became cold enough, they slowed down and there was no avoiding the hug of the Single Drops.

Once they got hugged and held tight by the Single Drops, the Very Small Drops stopped kicking about light. Light could finally fly in a straight line and the All-There-Is became like a clear glass of water.

The light from this very early moment has been on its way to us for a long, long time, and it has become very tired. But today we can still pick it up with our Far-Seers, and it tells us the story of the Single Drops, the Heavier Drops, and the Very Small Drops.

If we ask in the right way, this light can also tell us about what it has seen on its long way to us.

And one of the surprising stories of this tired light from the very beginning is that there is much more out there than we can actually see.

Most of what is out there is dark.

But what exactly this dark matter is, no one knows.

This is what took the student-woman to Big-Seer: she wanted to find out.

Part 5

Dark Rain

Her pink lips
Stay dry
In the heavy dark rain

What really brought her to Big-Seer is not what he can see with his big eye. Rather, it was all the other things that he can't see.

The dark matter.

It seems crazy that student-people should think that there is a lot more stuff that you can't see than stuff that you can see. Still, they do. In fact, they believe that there is about five times more dark matter than normal matter.

If you look around us with a Far-Seer, you will realize that the White Road is made of many, many stars. There are about four times ten stars in the White Road for each person on our Home-World.

And if you use a Big-Seer, you will find that there are as many other Star-Crowds in the sky as there are stars in the White Road.

And yet all of this is just a tiny bit of everything there is.

Doctor Zwicky was one of the first student-people to understand this. Many other student-people after him found that he was right.

At the time when many young people were growing long hair and asking for a better world without fights, a student-person called Doctor Rubin was the first woman to be allowed to use a Big-Seer.

She was following Mr. Zwicky's foot-steps. Mrs. Rubin's idea was that stars further away from the middle of a Star-Crowd should go around more slowly than those nearer the middle. This is because most of the stars that you can see are found in the middle of Star-Crowds. Their high numbers mean that this is where their pull is strongest and it becomes less strong the further out you go.

Mrs. Rubin reasoned that stars further out from the middle should slow down, or they would fly out because there isn't enough of a pull to keep them there.

But Mrs. Rubin used a Big-Seer and saw that in fact stars further out from the middle of Star-Crowds go around as fast as those nearer the middle. This was thought not to be possible, but it has been seen over and over again since then.

There are only two possible ways to understand this.

In the first way, the pull of stars on each other is different from what Mr. Newton and Mr. Einstein thought. This would mean that we need a new way of thinking about the pull between heavy things when they are far away from each other.

But then we would have to change Mr. Einstein's ideas about space-time. Yet we know he was right, for his ideas have been checked over many times.

The second way to understand what Mrs. Rubin found is to imagine that there is much more stuff out there than we can see. This stuff you can't see would have its own pull, on top of the pull given by the normal stars that you can see.

The pull coming from the dark stuff would be the reason why the stars far away from the middle can go around so fast: they are kept there by the pull of dark matter.

Suddenly, there was far more in the sky than anyone had imagined.

Since then, dark matter has been found at work in many places: pulling on stars inside Star-Crowds, changing the way the White Road turns around, and making far-away Star-Crowds dance with each other.

The more they searched for dark matter, the more screaming expressions of it the student-people found.

The pull of dark matter has even been found to change the way the All-There-Is grows.

Since dark matter is heavy—perhaps a few hundred times as heavy as a Single Drop, or more—its pull slows down the All-There-Is as it becomes bigger. We can tell that this happened by looking at far-away Star-Crowds and working out how much bigger the All-There-Is has become since the light we see left on its trip to us.

Dark matter also changes the form of space-time. Because of this, when light flies toward us from far-away Star-Crowds, it is dragged away from a straight line. Using Mr. Einstein's ideas, student-people can show that far-away Star-Crowds appear to us on the sky slightly less round than they should be.

It is as if light has a memory of the pull of dark matter it passed on its way to us.

Thanks to Big-Seer, the student-woman can now ask what that light coming from far-away Star-Crowds has seen on its trip.

She hopes that the answer will help explain what dark matter actually is.

Student-people have some bright ideas about what dark matter might be—but they are not sure yet whether they are right.

In the time it takes you to blink, the number of dark matter drops that fly through your hand is two times the number of people living today in the city that never sleeps.

They imagine that dark matter is a new type of drops, which are of a different kind from the matter you and I are made of.

Dark matter drops are thought to be part of a large group of drops called Mirror Drops—a type of drops no one has seen yet. Student-people believe that for each type of normal drop we know about, there should be a type of Mirror Drop that goes with it.

Mirror Drops are believed to be the same as normal drops in almost all ways, except for two: Mirror Drops are much heavier than normal drops and have a different relationship with their own kind. This would explain why we haven't found them yet.

Normal drops are grouped into two families. In one, drops love to hang around with drops of the same family and try to get together as much as they can. In the other, drops do not like having family around and do their best to keep themselves to themselves.

Mirror Drops also come in two families, except that their family relationships are the other way around: if a normal drop loves family, its Mirror Drop will avoid it; and the other way around.

This is like when you look at yourself in the mirror: what you see staring back looks almost the same as you, except it is flipped.

Mirror Drops could live early after the Big Flash, because everything was hot then and the heavy Mirror Drops could only come to life in hot places. But when everything cooled down, the Mirror Drops broke down into lighter and lighter drops, until only the lightest drop remained. This is what we call dark matter.

Dark matter is made up of the last remaining Mirror Drops from the Big Flash.

And it is all around us.

In the time it takes you to blink, the number of dark matter drops that fly through your hand is two times the number of people living today in the city that never sleeps.

This might be hard to believe. If there are so many dark matter drops, how come we do not feel them?

In fact, we spend our lives sitting in a dark matter rain, but we never get wet.

The normal drops we are made of do not feel dark matter drops most of the time. This is because dark matter drops can kick normal matter only slightly, and only once in a while.

Most of the time, dark matter flies through without leaving any trail.

But if we listen very carefully, we might be able to pick up the quiet whisper of dark matter.

Part 6

Whispers From the Dark

A soft song
Might tell you
Dark stories

She had come all the way up here to try and figure one thing out: whether dark matter was changing, ever so slightly, the direction of light coming from far-away Star-Crowds.

Some of her friends, people she went to college with, have become a different kind of student-person. They are trying to catch the whisper of the dark matter rain all around us.

And to do that, they need silence.

But not a normal kind of silence. They need to silence all other kinds of normal drops that would usually scream over the quiet dark matter drops.

To do so, they have to find a place where normal drops can't get to.

There are all kinds of normal drops flying around that you must take out, or else you could confuse their chuckle for the dark matter whisper.

Loud sounds come from bursts of fast drops coming from the sky. To take those out, student-people build dark matter ears deep inside rocks. Sometimes, they put the dark matter ears in deep mines, where other people bore to look for pretty things to put on their fingers.

Those are silent places, perfect for listening to dark matter.

Other confusing sounds come from a type of rock that we can use to power our computers, but also to form fire balls that can kill many people. Doctors use those rocks to help sick people when some of their cells won't stop growing.

To silence these rocks, student-people line the dark matter ear with lead, so that their screams remain outside.

Our Home-World sits in a dark matter rain coming from space, with most of the dark matter drops passing through without leaving any sign.

This dark matter rain is a bit stronger in the middle of the year than at the end because of the way our Home-World goes around the Sun. In the middle of the year, we go in the same direction as the dark matter rain, and against it six months later.

It is like walking in the rain on a day with lots of wind: if you walk against the wind, rain drops hit you harder than if you go in the wind's direction.

By putting a dark matter ear deep inside rocks, student-people try to catch the whisper of the dark matter drops flying through.

Because the dark matter rain changes with time, if you piece together all the dark matter whispers, you should get a song that changes gently from month to month and then repeats year after year.

Some student-people say that they have heard the change in the dark matter song during the year. But most other student-people do not believe them: other dark matter ears have not heard the same song at all, but just silence.

When student-people from different teams get together, there is always screaming and shouting, and people get mad at each other.

To sort this out once and for all, most student-people think that what we need to do is to build even bigger ears to listen more carefully.

Others think that the best way to see dark matter is to send a flying Far-Seer to the sky.

Dark matter does not give out light, but when two dark matter drops hug each other, they disappear. In doing so, they throw out other types of drops, which eventually give flashes of light.

But this light can't be seen by your eye. It can't reach us on the ground, either, because it is stopped by the air. Not even Big-Seer can see it.

The only way to catch it is to send up a Far-Seer able to see this kind of light.

The problem is that a lot of other things in the sky will give you the same kind of light, and those have nothing to do with dark matter.

Student-people came up with ways of blocking out the light from normal matter and look only at the flashes coming from dark matter hugs.

As with the dark matter song, some people say that they have caught light from dark matter hugs. But catching this light is a very hard business, and others are not sure the light is actually coming from dark matter.

The search continues.

Student-people also believe that dark matter should be found in the middle of the Sun, kept there because of how heavy the Sun is. The light from the dark matter hugs inside the Sun can't reach us, but some of the other drops given out by those hugs will escape and fly through space to us.

To pick up those drops, student-people have built a huge eye and put it inside the ice, in a cold place near the bottom of our Home-World. It's place where the Sun is always up for half the year, and always down for the other half. This eye is three times taller than the building of Mr. Eiffel!

Every now and then, the drops coming from dark matter hugs inside the Sun will kick normal drops inside the ice. The normal drops get kicked so hard that they are sent flying through the ice faster than light would go. As they slow down, they give out flashes of blue light. This is what the huge eye tries to see by looking at the ice all around it.

So far, the huge eye has seen many blue flashes, but none of them are from the dark matter hugs inside the Sun.

If it's so hard to pick up signs of dark matter around us, some student-people say, then perhaps what we should do is make some dark matter, catch it in a bottle, and study it there.

That would be really good. What we need to do is to make some normal drops very hot, just like they were right after the Big Flash, and then force them to hug each other. From their hugs, lots of other drops will fly out, and perhaps a few dark matter drops, as well. But we would need to get the normal drops crazy hot—at least a hundred times a hundred times a hundred times a hundred times a hundred times hotter than the inside of the Sun!

Many student-people from around the world have worked together for a long time to do this.

There is a city in a land full of safe places to put your money in. People in this land know how to make dark sweets that make your mouth water. They build tiny wood houses that tell the time with the song of a little flying animal, also made of wood. Mr. Calvin lived and worked in that city, once.

Near that city, student-people have built a large ring under the ground. It would take you over five hours to walk around that Big Ring.

Student-people take normal matter drops and make them fly around the Big Ring almost as fast as light. In a second, the drops go around the ring a hundred hundred times.

Then student-people pick a point where they make the normal matter drops hug each other, and they look at what kind of other drops come out of their hot kisses.

This way, student-people have already found a new type of drop that no one had seen before, but that a long time ago Doctor Higgs had said should be there.

Mr. Higgs was very happy about this.

Near that city, student-people have built a large ring under the ground. It would take you over five hours to walk around that Big Ring.

(This is one of the best things that can happen to you if you are a student-person: to say what something in the world should be like before anyone else has seen it, and then to find out that you are right. For this, Mr. Higgs and another student-man called Doctor Englert have been asked to go to the important party in the ice-water land.)

Besides the drop of Mr. Higgs, so far we have not seen any other types of new drops in the Big Ring.

Student-people hope that some Mirror Drops will be next.

All of those whispers from the dark keep pulling student-people forward in their search.

Everyone hopes she will be the one to see dark matter for the first time.

Many student-people who spend a lot of time drawing strange signs on paper believe that it should be possible to see dark matter in the next five years or so.

No one knows whether dark matter will finally show up in one of the big ears in the rock, in the flying Far-Seer in the sky, in the huge eye in the ice, or in the Big Ring in the ground.

But when it does, the news will be on television and will change the way we understand the world.

Part 7

Death by Fire

With their last breath
A blushing of light
From far-away stars

She steps outside in the cold night, holding her cup of hot coffee with both hands.

The White Road is beautiful in the dark, clear sky, and, once again, she can't help but be amazed by it all.

It does not matter how many times she has seen this before, or how much she knows about what is out there. The sight of the stars is enough to make her gasp.

"It all seems so still, and yet it's changing all the time," she whispers to no one.

It is hard to believe that everything out there past the White Road and its stars is running away from us.

Yet, like Mr. Hubble found long ago, the Star-Crowds are running away from each other, as the space between them gets bigger and bigger. The All-There-Is is growing with time.

She can still remember when the news came through that some student-people had found something very exciting about how fast the All-There-Is grows. She was in school then, and this was one of the things that had made her decide to become a student-woman.

The idea the student-people had used to figure out how the All-There-Is grows was quite easy.

Imagine you took a light from your night-table, placed it at the foot of the bed, and looked at it. The light would appear to be bright to you.

If you then moved the same night-table light to the end of the drive-way, it would seem much less bright.

If you now put it at the end of the road, the light would look even less bright.

So this means that the night-table light appears less bright to you the further away it is.

And here is the idea: if you look carefully at how bright the light appears, you can find out how far away it is.

That would work as long as you always used the same night-table light.

If you want to figure out how the All-There-Is is growing, what you need to do is to see how far away the Star-Crowds are and how fast they are moving.

Once you know these two things, you can work out how much the All-There-Is has grown between us and the far-away Star-Crowds.

Finding out how fast they are moving is easy. The light we get from Star-Crowds that are moving away faster is more tired. By looking at how tired the light is, we can figure out how fast the Star-Crowds are moving.

But finding out how far away the Star-Crowds are from us is much harder.

The reason is that Star-Crowds are all different: some are big, some are small. Some are old, some are young. Some are bright, some less so.

So by looking at how bright or how big Star-Crowds appear to us in the sky we can't tell whether they are close by and small, or far-away and large.

What student-people needed was something like your night-table light. If they could find a night-table light in each of the Star-Crowds, they could then figure out how far away they are.

But they needed a very bright night-table light, something that could be seen burning from the farthest corners of the All-There-Is.

Student-people found that the death of stars gave them the right night-table light.

Not just any star would do, though. What they needed was a kind of star that dies in a big show of fire and light that can be seen from far away.

They looked at a great number of star deaths, and found that some Dying Stars could be used like night-table lights, because with their last breath they all gave out nearly the same light. The student-people weren't sure why, though. Perhaps it was because these Dying Stars are all made of the same stuff and all die when they become heavier by eating up another star.

No one knows exactly why it works. But it does.

The student-people managed to catch a few Dying Stars in different Star-Crowds.

Thanks to those, they worked out how far away the Star-Crowds are. They also knew from the tired light how fast the Star-Crowds are moving.

And when they put those two things together, they found something no one expected.

Student-people found that the death of stars gave them the right night-table light.

Part 8

The Dark Push

A silent end
Times of light
Half remembered

The Dying Stars told a weird story to those who found them.

As student-people expected, the light from the far-away Dying Stars was less bright than the light from those near us—just like the night-table light appears brighter the nearer it is to you.

The big surprise was that the far-away Dying Stars appeared even less bright than they should, given how tired their light was. There was no way those Dying Stars could be so far away from us, except if they had been pushed away by some strange force.

But it was also possible that part of the light from the far-away Dying Stars had been somehow simply taken away during its trip to us. The light could have passed through groups of matter drops that blocked it out, and in this way make the far-away Dying Stars look less bright than they should.

But the student-people checked all this very carefully and could not explain it this way.

Or it could be that the far-away Dying Stars are different from the ones near us. They could be less bright to begin with, which would make us think they are further away than they are. That would be like using a different night-table light when you put it at the end of the drive-way.

But again, no one could find anything to show this to be true.

There remains only one way to explain it. And the key idea is Mr. Einstein's Dark Push.

Before Mr. Hubble's work, Mr. Einstein had believed that the All-There-Is was not changing with time. Yet he knew that the pull of the matter drops all around us would make the All-There-Is change.

But this he could not accept. So he made up something new, something that would exactly counter the pull of matter: the Dark Push. Such a Dark Push was a strange idea. Instead of pulling heavy things closer to each other, like any normal matter would do, the Dark Push was meant to push things away from each other. Big, heavy things—as big as the Star-Crowds we see in the sky.

All other Star-Crowds will fly away from us faster and faster until at some point, far away in time, we won't be able to see them anymore.

One thing was clear: the Dark Push could not be anything like the kind of matter or forces we know about. It had to come out of no-where—perhaps out of empty space.

What if all that dark space was not actually empty?

If matter drops came into life and died very quickly all the time, we would not be able to notice it. But this breathing in and out of drops by empty space would change the way the All-There-Is grows. After all, there is a lot of empty space in the All-There-Is!

If empty space could breathe in and out just enough drops—not too many, not too few—Mr. Einstein could have his never-changing All-There-Is, just as he wanted.

Mr. Hubble then showed that Mr. Einstein was wrong, and everyone dropped the Dark Push. But then everything changed again because of the big surprise from the Dying Stars.

The Dying Stars were saying not only that the All-There-Is is growing with time, but that in fact it is growing faster and faster with time!

Nothing could do that, except Mr. Einstein's Dark Push.

So Mr. Einstein had been right, after all, if for the wrong reasons.

He wanted the Dark Push to stop the All-There-Is from changing. A hundred years later, the Dying Stars have shown student-people that the Dark Push is actually making the All-There-Is grow faster and faster all the time!

"The more we learn about the All-There-Is, the more questions we are left with," she sighs.

The student-people already had a deep question to answer: what is dark matter made of? Now they had two more questions: where does the Dark Push come from? And why do the two together add up to twenty times more than the normal matter out there?

Until those questions are answered, many student-people will spend many nights without sleeping, thinking about them.

One thing appears to be clear. If the Dark Push is what the Dying Stars say it is, then it will continue to push things away from one another.

The bigger the All-There-Is becomes, the more empty space there will be and the more Dark Push. This will make the All-There-Is grow even bigger, even faster.

And so on. For all time.

You can't stop the Dark Push.

All other Star-Crowds will fly away from us faster and faster until at some point, far away in time, we won't be able to see them anymore.

They will be too far away for their light to reach us. Stars will die, and the Star-Crowds will turn off.

The All-There-Is will become even bigger, and almost totally empty. Nothing will be left but dark silence.

Part 9

Is the All-There-Is All There Is?

So many worlds
All that can be
Is

Her hands have become cold and she goes back inside. She closes the door behind her and goes past the room where Big-Seer is going about his business.

She sits down in front of her many computers. Everything seems to be going fine. She has another couple of hours before the night is over.

She asks Big-Seer to move his eye to the next set of Star-Crowds, and he does so quietly.

As she writes down a few numbers, a weird thought makes her pause: perhaps somewhere out there, another woman who looks exactly like her, who is just like her in everything but the smallest thing, is doing exactly the same job right now.

It is even weirder that this could have something to do with the Dark Push.

She has thought those thoughts before. Still, they always make her head race.

Because for the student-people who are trying to understand the All-There-Is, the All-There-Is is not big enough. Not if they want to explain where the Dark Push comes from.

It all starts with a question, another one of those hard Why? questions: why is the Dark Push as strong as it is, and not a little bit different?

It is not clear at first why this is an important question. It is not even clear that it is a question that makes sense at all!

But imagine for a minute the following situation.

You enter a room where you find a table with a large number of small, gray, round pieces on it—of the type that you can use to buy a coffee, or a paper, or to pay for parking. The ones with one head on one side and some other picture on the flip side.

Let's say that there are four hundred of the gray pieces on the table. And they all show heads.

Let's say that there are four hundred of the gray pieces on the table.
And they all show heads.

You would not believe for a second that they were all just thrown on the table and happened to land this way. Although this could happen, it would be a hard thing to accept.

It would be easier to imagine that someone had walked into the room before you and had put them all down like this, heads up, all four hundred of them.

The strange thing about the Dark Push is that it is a bit like the four hundred heads-up gray pieces in the room.

If the Dark Push were only a tiny bit larger than it is, then everything we see around us would be very different.

It is as if changing only one of the heads in the four hundred would make the entire world change.

Change the Dark Push by a little bit, and Star-Crowds could not form; none of the stars we see in the sky would be there; the Sun would not be there; our Home-World would not be there; and life, as we know it, could not be here.

We wouldn't be here to talk about this in the first place.

So the question is: Who or what put down all four hundred heads exactly this way?

Some student-people came to believe that they could understand this by imagining more rooms. A very large number of rooms.

In each of them, the four hundred gray pieces are all thrown up in the air and flipped. And they land in some way, however they may.

In most of the rooms, some of pieces will land heads, and some won't.

But if you have enough rooms, in the end you'll find one room where all of the pieces have landed heads-up. Just like that.

There is no need to imagine anyone setting them up in this way.

It's only a question of having enough rooms and trying them all.

And so the idea is that perhaps the All-There-Is is not all there is. This is what made her head race.

If we had many kinds of All-There-Is, all different from each other in some way, then one of them will be exactly right, with all the four hundred gray pieces showing heads, with exactly the right Dark Push in it for Star-Crowds to form, our White Road to form, the Sun to form, for us to live there and to be asking those hard Why? questions!

Maybe there are so many different kinds of All-There-Is out there, that you could not even number them all.

Most of them would be empty of life, dead and boring. But some of them will be just right for us to live in.

And some will have a person in there who looks just like you. Maybe a tiny bit different from you: a different hair color, different eyes, or wearing a different shirt.

To the student-woman, this sounds more than a little crazy.

And yet, many times before, very crazy ideas have turned out to be right, after all: Mr. Hubble and his flying-away White Shadows; Mr. Einstein and his space-time; the dark matter and the Dark Push.

All of those things remind us that the All-There-Is can sometimes be more crazy than our craziest ideas.

But perhaps the most amazing thing about the All-There-Is is that we can understand it at all. Even though we still have many very big, very hard Why? questions, there are lots and lots of things we understand and can explain about everything around us.

The All-There-Is speaks in a tongue that student-people have learned to understand, little by little.

But there is still so much more left to tell.

Part 10

When the Sun Comes Up

Everything opens
At the smallest touch
Of a rose

She sits down. The big blue body of water in front of her seems to go on without edges and without end.

She can feel the warm hand of the Sun on her face.

She feels happy.

The night's work has gone well. Big-Seer has done a great job, the best that could be done. She can go home now.

But her job has only just begun. There is much more left to do in the coming weeks and months before she can make sense of what Big-Seer saw last night.

She is looking forward to it.

Letters and words and entire books are hidden in what Big-Seer has given her, written in the strange tongue of the All-There-Is.

Little by little she will understand it better and better.

All she needs to do is ask the right questions in the right way, and she might learn the truth.

She smiles, and the Sun smiles back at her.

The big blue body of water in front of her seems to go on without edges and without end.

Thanks

I would like to thank the many people who helped me make this slightly crazy idea real.

Thanks to Randall Munroe for coming up with the idea of using the most-used ten hundred words in his drawing of the Up-Goer Five.

To my student-people friends, Gianfranco Bertone, Dave Clements, and Jennifer Siegal-Gaskins, thank you for helping to make this book better with your clear tongue. To Gianfranco, thank you for those many coffees under a totally blue sky and for all the things, big and small, we talked about.

To Ed Dark and Stephen Follows—two guys who write and make funny movies—thank you for seconding this idea early on, and for laughing out loud when I read some bits out to you. It made me feel like this could really fly.

To my friends Terry and Dena Higgins, thank you for all your comments and for believing in this.

To Peter Tallack—a guy people like me turn to in order to get their books read by as many people as possible—thank you for saying yes straight-away and for not giving up.

To T. J. Kelleher—a guy who says yes or no to would-be writers—thank you for taking this on.

To Tisse Takagi—a woman who checks books very carefully to spot if anything is wrong or can be written better—thank you for questioning everything.

To all the people working with T. J. who have helped in putting this book together, many thanks for doing such a great job.

To Antoine Déprez, thank you for the beautiful pictures that make this book so much nicer.

To my daughter, Emma, thank you for putting a smile in my heart every time I look at you.

To my wife, Elisa, thank you for making sure every line was as clear as it could be. Most of all, thank you for being the brightest star in my All-There-Is. Without you, none of this would be here.

This book is made of seven hundred and seven different words and one short of two times twenty names of people.

Some Expressions Explained

All-There-Is: **universe.**

Big Flash: **Big Bang.** The English astronomer Fred Hoyle coined the term *Big Bang* in 1949 to describe the primordial explosion that gave rise to our universe. Hoyle himself did not believe that the Big Bang was the correct theory for the origin of the universe, but the name stuck. We now have overwhelmingly strong evidence that the Big Bang happened 13.798 billion years ago. The margin of error on the age of the universe is a mere 37 million years.

Big Ring: **Large Hadron Collider (LHC).** The largest and most powerful particle accelerator in the world, the LHC is a huge, underground, ring-shaped machine, twenty-seven kilometers in circumference. It is located at the CERN laboratory, near Geneva, Switzerland. By accelerating and colliding protons (a type of elementary particle) at very high energy, it seeks to produce new types of particles, which might include dark matter. The Higgs boson was discovered there in March 2013.

Big-Seer: **large telescope.** The world's largest optical telescopes are built on top of high mountains so as to minimize the amount of atmosphere above them—which introduces distortions to the images—and to avoid light pollution. The largest telescopes today have mirrors ten meters in diameter and sophisticated, computer-controlled "adaptive optics" mechanisms that compensate for the residual fluctuations in the air and thus deliver a sharper image of distant objects.

Crazy Stars: **planets.** Our solar system is made of eight planets orbiting around the Sun. A ninth object, called Pluto—which traditionally was considered a planet despite its small size and large distance from the Sun—was reclassified as a *minor planet* by the International Astronomical Union in 2006.

Dark Push: **dark energy.** Looking for a solution to his equations that would describe a static universe (one that did not expand or contract with time), Einstein invented the *cosmological constant*—a repulsive force arising from empty space that could counteract exactly the pull of gravity, thereby making the universe unchanging with time. Today, observations of explosions of distant stars called *supernovae type Ia* have demonstrated that something similar to Einstein's cosmological constant makes up about 70 percent of the contents of our universe. We call this mysterious, repulsive force *dark energy*. Its origin is presently unknown.

drop of Mr. Higgs: **Higgs boson.** In 2013, the ATLAS and CMS experiments at CERN made headlines all around the world by discovering a new type of elementary particle—the Higgs boson. First predicted to exist by British physicist Peter Higgs and others in the 1960s (a feat for which Higgs and François Englert shared the Nobel Prize for physics in 2013), the Higgs boson is the particle that gives mass to all other elementary particles.

Dying Stars: **exploding stars.** When certain types of stars reach the end of their life, they explode in what is called a *supernova*, emitting an enormous amount of light and energy. For a few days, a single supernova can become brighter than an entire galaxy! *Type Ia supernovae* are a particular class of such explosions, characterized by their almost uniform brightness. This uniformity makes them suitable as "standard candles," which astronomers can use to measure distances in the universe. The Nobel Prize for physics in 2011 was awarded for studies of supernovae type Ia, which led to the conclusion that the universe contains a large amount of dark energy.

Early Push: **cosmological inflation.** The light left over from the Big Bang (the so-called *cosmic microwave background*) has almost the same temperature, independent of where you look on the sky, with only tiny fluctuations in different directions. This points to the hot beginnings of our universe and also tells us that the size of the visible universe increased by at least 10^{26} times in the first 10^{-32} seconds after the Big Bang. This incredibly quick expansion, called *cosmological inflation,* is believed to have been powered by an as-yet undiscovered new type of energy field. In March 2014, the BICEP2 telescope claimed to have observed from the South Pole the gravitational waves imprinted by inflation onto the cosmic microwave background—a discovery that, if confirmed, would be definitive proof of the theory of inflation.

far-away Crazy Stars: **exoplanets.** Until 1992, no other planet outside our solar system was known. Astronomers have since developed new observational techniques that allow them to "see" the effect of planets orbiting distant stars. We now believe that most stars in our galaxy have planetary systems, and over seventeen hundred exoplanets have been discovered to date. So far, we haven't been able to locate Earth-like exoplanets, but as observational techniques improve, such a discovery is only a matter of time.

Far-Seer: **telescope.**

Fight-God: **Mars.** With its distinct red, almost bloodlike color, Mars was named after the Roman god of war. Many unmanned missions have investigated the Red Planet, and some have landed computerized rovers that have sent back spectacular pictures of what is now a desolate desert. There is evidence that Mars once had flowing water, which might still survive today in ice form.

Great-Father-God: **Saturn.** Saturn was the Roman equivalent of the Greek god Chronos, who was ousted as head of the gods by his son Zeus (Jupiter for the Romans). The planet is characterized by spectacular rings made of tiny ice particles and debris. Saturn is a gas giant, meaning that it is not composed primarily of rock but is actually made of gases (mostly hydrogen and helium). Saturn has no firm surface you can stand on. The planet has several dozen moons, including Titan.

Head-God: **Jupiter.** The largest planet in the solar system, Jupiter is a gas giant like Saturn. Seen through a small telescope, it reveals a red spot (the so-called Great Red Spot) on its surface, which is thought to be a huge storm system (two or three planets the size of the Earth could fit inside it!). Jupiter has several moons, the largest four of which are called *Galilean moons*, as they were first observed by the Italian astronomer Galileo Galilei in 1610. The planet is named after the father of the gods (who was also their king) in Roman mythology.

Heavier Drop: **helium.** The nucleus of a helium atom is made of two protons (positively charged particles) and two neutrons (electrically neutral particles). We believe that most of the helium in the universe was formed in the first three minutes after the Big Bang. Helium is the second-most abundant element in the universe, after hydrogen.

He-Who-Talks-for-the-Gods: **Mercury.** In Roman mythology, Mercury was the messenger of the Gods. He was depicted as a winged young boy, holding a short staff entwined by two snakes, called *caduceus*. Mercury is the first planet from the Sun. Its year is only 116 days long, compared with Earth's 365, which might be the reason why Mercury was associated with the fast-flying messenger of the Gods.

Home-World: **Earth.** Earth is the third planet from the Sun, and the only one that supports life, as far as it is known.

Mirror Drops: **supersymmetric particles.** Particle physicists think that for every type of known elementary particle, there should exist another particle (*supersymmetric particle*) with almost exactly the same properties. A key difference is that one of the fundamental properties of particles (their *spin statistics*) is reversed in the corresponding supersymmetric particle—a bit as if the particle was seen in a mirror. Supersymmetric particles also have a much larger mass than ordinary particles. Although presently unobserved, supersymmetric particles would solve many problems of our current understanding of particle physics. They would be abundant in the universe right after the Big Bang, but then quickly disappear, leaving behind only the lightest of them. This relic supersymmetric particle could be the dark matter in the universe—which we believe constitutes about 25 percent of the matter-energy content of the cosmos today. It is hoped that supersymmetric particles will be discovered at the LHC.

normal drops: **elementary particles.** The matter all around us is made of fundamental building blocks, each with its own properties such as mass, electric charge, and other, more exotic ones. Ordinary matter is made of protons (positively charged particles), neutrons (neutral particles), and electrons (negatively charged particles). But there are many other types of elementary particles. Other particles are associated with forces. For example, the photon (i.e., the particle responsible for light) is the particle that carries the electromagnetic force.

She-God of Love: **Venus.** Venus is the nearest planet to Earth, a mere thirty-eight million kilometers away when at its closest. It's called the *morning star* or *evening star,* as it becomes brightest shortly before sunrise and shortly after sunset. It is named after the Roman goddess of beauty and love.

Single Drop: **hydrogen.** A hydrogen nucleus is made of a single positively charged particle, the proton. Hydrogen is the most abundant element in the universe, and the simplest one.

Sister Drops: **antimatter particles.** Antimatter particles have the same properties as their matter counterpart, but opposite electric charge. Antimatter and matter are like twin sisters, but with a crucial difference. When a matter particle collides with its antimatter counterpart, the two particles disappear and their energy and mass are converted into other particles (including light). It is an unsolved mystery why the universe appears composed of matter rather than antimatter. The evidence is that a tiny, unexplained excess of matter over antimatter (one part in a billion) in the primordial universe led to the survival of the excess matter particles, while every other matter-antimatter pair disappeared.

Star-Crowd: **galaxy.** There are many types of galaxies in the universe, from beautiful spiral galaxies to cigar-shaped elliptical galaxies, from dwarf galaxies to irregular ones. A typical galaxy contains about a hundred billion stars, a great deal of gas, and even more dark matter, in the form of an invisible halo that surrounds it.

Student-people: **scientists.**

Sun's Sister: **Moon.** The Moon is so far the only other place in the solar system humans have set foot on, apart from the Earth. We are still unsure how the Moon was formed, and whether it split out from the Earth as a consequence of a giant impact or was captured by gravitational attraction. By a lucky coincidence, the apparent size of the Moon in the sky is almost the same as the Sun's (although the Sun is four hundred times further away from us than the Moon). This means that we get treated to spectacular total solar eclipses, which wouldn't happen if the apparent size of the Moon were smaller than the Sun's.

Very Small Drops: **electrons.** A type of elementary particle carrying negative electric charge. Electrons orbit around atomic nuclei, making atoms overall neutral. In the early universe, right after the Big Bang, the temperature was too high for electrons to be able to orbit around a nucleus, so they were dislodged and whizzed around free. In so doing, they scattered light particles, which made it impossible for light to travel very far. The early universe was thus an opaque plasma. Once the temperature cooled sufficiently, 380,000 years after the Big Bang, electrons were captured by hydrogen nuclei and the universe became transparent.

White Road: **Milky Way.** Our own galaxy, the Milky Way, is made of a few hundred billion stars. It can be seen as a fuzzy white strip across the sky on a clear, dark night—hence its name. The solar system completes an orbit around the center of the Milky Way every 250 million years.

White Shadows: **distant galaxies.** We now know that the fuzzy, faint objects that astronomers talked about as "nebulae" a hundred year ago are nothing but distant galaxies. It is estimated that there are over a hundred billion galaxies in the observable universe.